STRATEGY FOR WATER QUALITY STANDARDS AND CRITERIA

OFFICE OF SCIENCE AND TECHNOLOGY

Setting Priorities to
Strengthen the Foundation for
Protecting and Restoring the Nation's Waters

Foreword

Water quality standards and criteria are the foundation for a wide range of programs under the Clean Water Act. This strategy contains priority strategic actions that the Office of Science and Technology will undertake in collaboration with other EPA offices, states and authorized tribes over the next six years to strengthen and improve this foundation.

While developing this strategy, we were frequently reminded of the importance of a strong standards and criteria program. Beginning in 2001, we held extensive listening sessions and frank discussions with states, other partners and EPA staff to obtain information, views, and ideas about needs for the water quality standards and criteria program. We also considered the recommendations regarding standards and criteria in the National Research Council's 2001 report, *Assessing the TMDL Approach to Water Quality Management*, and from the General Accounting Office's 2002 report, *Improved EPA Guidance and Support Can Help States Develop Standards That Better Target Cleanup Efforts*. This strategy is designed to carry out our mission under the Clean Water Act, to address the needs expressed by our partners, and to support EPA's Strategic Plan. It also includes many changes in response to public comments on the May 2002 draft.

The Office of Science and Technology will continue to work with its partners as we implement the strategy, and some priority actions are already well underway. We will continue our dialogue with partners as we track progress and adjust efforts each year to stay on the strategic course we have set. Progress will be reported at http://www.epa.gov/waterscience/standards.

Geoffrey H. Grubbs

Geoffrey H. Grubbs, Director
Office of Science and Technology

Contents

Disclaimer

The discussion in this document entitled "Strategy for Water Quality Standards and Criteria: Setting Priorities to Strengthen the Foundation for Protecting and Restoring the Nation's Waters" is intended solely as a planning document for the Office of Science and Technology within EPA's Office of Water. The statutory provisions and EPA regulations described in this document contain legally binding requirements. This strategy is not a regulation itself, nor does it change or substitute for those provisions and regulations. Thus, it does not impose legally binding requirements on EPA, states, tribes, or the regulated community. This strategy does not confer legal rights or impose legal obligations upon any member of the public.

While we have made every effort to ensure the accuracy of the discussion in this strategy, the obligations of the regulated community are determined by statutes, regulations, or other legally binding requirements. In the event of a conflict between the discussion in this strategy and any statute or regulation, this document would not be controlling.

This is a living document and may be revised periodically without public notice. We welcome public input on this document at any time. The general descriptions provided here may not apply to a particular situation based upon the circumstances. Interested parties are free to raise questions and objections about the substance of this document and the appropriateness of the application of this document to a particular situation. EPA and other decision makers retain the discretion to adopt approaches on a case-by-case basis that differ from those described in this document where appropriate.

Executive Summary

Water quality standards and criteria are the regulatory and scientific foundation of programs established under the Clean Water Act to protect the Nation's waters. As such, they are among the most critical clean water programs. Due to the many new demands on the program, and since the nature of water pollution problems and required solutions have changed dramatically in recent years, water quality standards and criteria need to be made a high priority and given a renewed focus. The water quality standards and criteria program needs clear priorities to address these critical demands.

This strategy is the product of a wide-ranging review of the existing water quality standards and criteria program within the context of all clean water programs. The review covered clean water goals, mandates and authorities; EPA's current strategic goals for clean water and other strategic planning efforts; and major needs of the current EPA standards and criteria program and key programs linked to it including water quality monitoring, total maximum daily loads (TMDLs), National Pollutant Discharge Elimination System (NPDES) permits, nonpoint source programs, oceans and wetlands programs, and source water protection. The review considered the results of more than 50 listening sessions with over 350 people during April-September 2001 and recent recommendations from the National Research Council, the General Accounting Office, and EPA's Inspector General.

This strategy is built upon a long-term vision for the future:

All waters of the United States will have water quality standards that include the highest attainable uses, combined with water quality criteria that reflect the current and evolving body of scientific information to protect those uses. Further, standards will have well-defined means for implementation through Clean Water Act programs.

States support this long-term vision and look to EPA to help fulfill it, beginning with the priority strategic actions contained in this strategy. These actions in the strategy are designed to address the following strategic themes:

- Filling major program gaps to achieve critical environmental results. For example, the water quality standards and criteria program needs to help states strengthen water quality criteria for three pollutants (sedimentation, pathogens, and nutrients) that are responsible for an estimated 40 percent of water quality impairments nationally.

- Strengthening and maintaining the scientific foundation of water quality programs, including targeting criteria development for specific pollutants of highest importance.

- Clarifying for states how to implement key scientific and technical components of standards and criteria when regulating discharges.

- Establishing important technical and policy linkages between the water quality standards and criteria program and other programs such as those that protect drinking water.

- Broadening participation in the water quality standards and criteria program with states and other stakeholders.

The strategy describes and sets milestones for the ten strategic actions of highest priority for addressing these findings. These ten highest priority strategic actions are:

1 Issue implementation guidance for the 1986 bacteria criteria for recreation.	**7** Provide technical support, outreach, training and workshops to assist states and tribes with designated uses, including use attainability analyses and tiered aquatic life uses.
2 Produce and implement a strategy for the development of pathogen criteria for drinking water and recreational use.	
3 Produce and implement a strategy for the development of suspended and bedded sediment criteria.	**8** Provide implementation support concerning technical issues affecting permits and TMDLs, beginning with technical support and outreach concerning the duration and frequency component of existing water quality criteria.
4 Provide technical support to states and tribes for developing and adopting nutrient criteria and biological criteria.	
5 Develop and apply a systematic selection process to produce new and revised water quality criteria for chemicals to address emerging needs.	**9** Identify any drinking-water source waters whose water quality standards do not protect the use, and work with regions, states, and tribes to correct any deficient standards as soon as possible.
6 Complete the national consultation with the Federal Services on existing aquatic life criteria.	**10** Develop a web-based clearinghouse for exchanging information on critical water quality standards issues, beginning with antidegradation.

The strategy also contains five strategic actions, outlined in Chapter 2, which are next in priority for implementation. The Office of Science and Technology (OST) in EPA's Office of Water will work closely with other EPA programs, states, authorized tribes and stakeholders to implement the strategy.

Developing the Strategy

Background

EPA's water quality standards and criteria program supports and oversees the efforts of states and authorized tribes to set water quality standards for all waters of the United States. Water quality standards—consisting of designated uses for waters, water quality criteria to protect the uses, and antidegradation policies—serve the dual purposes of establishing water quality goals for specific water bodies and providing the regulatory basis for establishing certain treatment controls and strategies. EPA provides policy guidance and the latest scientific information to help states and tribes adopt standards. The Clean Water Act also requires EPA to review new and revised standards, approve or disapprove them, and issue federal replacement standards to correct deficiencies where necessary. The Office of Science and Technology (OST) in EPA's Office of Water (OW) is the headquarters office responsible for these efforts while the ten EPA regional offices have the lead for working with states and authorized tribes.

The Role of Standards and Criteria in Water Quality Programs

Water quality standards and criteria are undeniably key to protecting the quality of our Nation's waters. Water quality standards establish the environmental baselines used for measuring the success of Clean Water Act programs. In an evolving scientific arena, adequate protection of fish and wildlife, recreational uses, and sources of drinking water depends on having well-crafted standards and criteria in place for our waters. Having clear numeric baselines is also important for establishing treatment controls; for conducting watershed planning, protection and restoration; and for innovations such as market-based incentives and trading.

Most states developed water quality standards and criteria on a significant scale in the 1970s when the water quality problems being addressed were simpler: for example, assuring adequate dissolved oxygen for fish and shellfish and installing wastewater treatment systems for basic sanitation. These standards and criteria were rarely fine-tuned to address complex issues

such as protecting endangered or threatened species, addressing sedimentation and flow, addressing pathogens, evaluating site-specific attainability, or evaluating cumulative effects from combinations of pollutants or stressors.

For several decades EPA and states focused more on technology-based controls than on water quality-based programs such as water quality standards. The most recent focus on TMDLs, in some cases under challenging deadlines, and on resolving complex NPDES permit issues, has heightened the immediate need to strengthen the standards program in many areas. With EPA's assistance, states and authorized tribes have reviewed and updated these standards on an ongoing basis; however, evolving science, dramatically increasing implementation demands, and other circumstances have often significantly outpaced these efforts. Examples of evolving science include the need to update criteria based on new information, the need to reflect newly-understood local variations in pollutant chemistry and biology, the need for clarity in the implementation of new and existing criteria, and the desirability of having more direct measures of designated use protection through biological criteria.

As the Nation has grown over the past 30 years, so too has the complexity of water quality problems. States, tribes, and EPA need a common understanding of how to implement standards and criteria provisions when monitoring and assessing water quality and developing NPDES permits, TMDLs and nonpoint source controls. For example, states, tribes and EPA should have similar approaches for determining which waters are in attainment, setting designated uses, translating narrative criteria into numeric values, establishing mixing zones, or allowing variances to standards.

Given the increasing number and complexity of water quality standards issues that must be addressed, EPA, states and tribes need to partner strategically to address them in a way that will best resolve the most critical issues and ensure the protection and restoration of our waters.

Partnerships to Improve the Program

OST met with many partners inside and outside the Agency who depend on water quality standards to help identify the key challenges faced by the water quality standards and criteria program. We found that all partners are facing a daunting and complex workload to meet these challenges, and are looking to EPA for leadership. But we also found that all partners share the desire to improve the program and are willing to work with us. We are confident that by working together we can address the highest priorities among the growing list of short- and long-term needs and help achieve our goals for safe and clean water.

Scope of This Strategy

This strategy focuses on what OST and other EPA offices need to accomplish to meet the program needs of EPA, the states and authorized tribes. In this document, "states" generally means the state, territorial and interstate agencies that have water pollution control responsibilities. "Authorized tribes" means federally-recognized Indian tribes for which EPA has given approval to administer water quality standards programs. For Indian country as a whole, the strategy supplements but does not replace the goals and objectives for water quality standards expressed in the Agency's Strategic Plan and in *Protecting Public Health and Water Resources in Indian Country: A Strategy for EPA/Tribal Partnership*, EPA Office of Water, October 1998.

EPA has specific statutory and regulatory obligations under the Clean Water Act, including reviewing new and revised standards, approving or disapproving them, and issuing federal replacement standards to correct deficiencies where necessary. EPA also has obligations under other statutes, such as the Endangered Species Act. Additionally, from time to time EPA receives judicial mandates, enters into settlement agreements, or becomes subject to specific Congressional requirements. EPA takes all of these obligations seriously, and carries them out on a daily basis. OST did not list these responsibilities as actions under this strategy because they are not optional and hence not subject to priority-setting. The presence or absence of actions in this strategy should not be construed as altering our basic responsibilities.

How This Strategy Was Developed

The strategy is the product of a wide-ranging review and analysis of the water quality standards and criteria program within the context of all clean water programs. It was developed by a work group chaired by OST staff. The group first developed a draft list of issues addressing the major needs of the standards and criteria program and of programs that link to water quality standards, including water quality monitoring and assessment programs, the TMDL program, the NPDES permit program, the wetlands and dredge and fill permit programs, ocean protection

---

EPA IS HELPING MORE TRIBES TO RUN WATER QUALITY STANDARDS PROGRAMS

▶ The Office of Water's October 1998 strategy for Indian country sets a goal that "by 2005, 15% of tribes will have final water quality standards approved by EPA for waters under their jurisdiction." It calls for EPA to provide guidance, technical assistance, training, outreach, and workshops for interested tribes to set up and run standards programs.

▶ This work has paid off: since 1998 the number of tribes with standards has increased by 50%, from 14 to 21, making it the largest non-grant tribal program in EPA. Nevertheless, it is still only 4% of all tribes. Tribes face many technical and administrative challenges in establishing standards.

▶ OST and EPA's regional offices are continuing to implement the 1998 strategy to assist tribes, including considering the establishment of federal water quality standards for waters in Indian country that do not have standards.

programs under the Clean Water Act, and the source water protection rogram under the Safe Drinking Water Act. The workgroup then conducted more than 50 listening sessions with over 350 people during April-September 2001. Appendix 3 lists the information sources for this strategy, including the groups who participated. The listening sessions gave participants an opportunity to identify the most important issues for stakeholders and the timing of their needs regarding water quality standards. They also helped elucidate barriers and define emerging challenges. The workgroup also considered the following:

- Clean Water Act goals, mandates and authorities that pertain to water quality standards and criteria, including EPA's oversight responsibilities under section 303(c) of the Act and EPA's scientific information responsibilities under section 304(a) of the Act.

- The strategic goal for safe and clean water, together with objectives and subobjectives in EPA's Strategic Plan.

- Public comments and statements in public meetings in response to the 1998 Advance Notice of Proposed Rulemaking on the water quality standards regulation.

During this time the National Research Council (NRC) of the National Academy of Sciences issued a report, *Assessing the TMDL Approach to Water Quality Management*. It contained several major recommendations concerning water quality standards and criteria *(see box at left)*. These recommendations played a strong role in shaping this strategy. OST also considered recommendations from the General Accounting Office and EPA's Inspector General in separate studies listed in Attachment 1.

Additionally, this strategy is built upon a long-term vision for the future of water quality standards and criteria. This vision statement is the essential mission of the water quality standards and criteria program.

VISION

All waters of the United States will have water quality standards that include the highest attainable uses, combined with water quality criteria that reflect the current and evolving body of scientific information to protect those uses. Further, standards will have well-defined means for implementation through Clean Water Act programs.

States support this long-term vision, and look to EPA to help fulfill it, beginning with the strategic actions contained in this strategy.

Strategic Themes

This strategy prioritizes actions that EPA will lead, often in conjunction with its implementing partners and affected stakeholders, to improve water quality. The strategy contains priority strategic actions that OST believes are critical to tackle the most important environmental problems, accelerate the adoption and use of appropriate water quality standards, reduce the burdens and impediments to program implementation, and promote broad participation in activities affecting the Nation's receiving waters.

The review and analysis provided a fresh look at all aspects of the current water quality standards and criteria program. It resulted in ten strategic actions representing the highest priority in the strategy, and five strategic actions representing the next set of priorities that will guide OST activities in the coming years. These actions will help EPA in addressing the following major findings:

- Filling major program gaps to achieve critical environmental results.

  - For example, the water quality standards and criteria program needs to help states strengthen water quality criteria for three pollutants (sedimentation, pathogens, and nutrients) responsible for an estimated 40 percent of water quality impairments nationally.

- Clarifying for states how to implement key scientific and technical components of standards and criteria when regulating discharges.

  - For example, water quality criteria documents published in the 1980s contained detailed scientific information used for deriving criteria values but little in the way of guidance on how to interpret them when assessing attainment.

- Establishing important technical and policy linkages between the water quality standards and criteria program and other programs such as those that protect drinking water.

- Broadening participation in the water quality standards and criteria program with states and other stakeholders.

- Strengthening and maintaining the scientific foundation of water quality programs.

Setting Priorities

OST has narrowed the focus of the final strategy to ten highest priorities and five next priorities as mentioned above. In setting priorities, OST considered such questions as: Would the action provide an important link for restoring and maintaining the Nation's water quality? Would it be critical for meeting the Agency's goals for clean and safe water? Would it meet a critical need of states and tribes? Would it meet a critical need of a related water quality program such as monitoring, assessment, TMDLs, permits, or source water protection? Would the action address a major gap or lack of clarity in the existing EPA standards and criteria program? Would the action address the increasing scientific and policy complexities posed by the accelerating pace of efforts to restore impaired water quality? Would it respond to one or more of the five strategic themes listed above?

In the past few months, OST has shared drafts of these priorities with the workgroup and other EPA offices and made modifications on an iterative basis. We are confident that the final priorities have solid support among those who depend on these products the most.

This strategy does not include a priority strategic action to revise the national water quality standards regulation to address any implementation issues. OST believes that a revised regulation would not be the best way to address most of the issues raised during listening sessions. Most such issues derive from lack of clarity for implementing existing requirements, not because of defects in the regulatory requirements themselves. During listening sessions, participants generally suggested how EPA can address important implementation issues with policy and guidance. Specific issues may emerge in the future that can best be resolved by establishing new or revised national regulatory requirements, but such steps at this time are not warranted. Additionally, EPA is currently reviewing petitions received from interest groups to revise its regulations in certain areas. EPA has not yet completed its review of these petitions. If EPA decides revised regulations are necessary, OST will modify this strategy accordingly.

Priority Strategic Actions

Highest Priorities

The ten strategic actions below have the highest priority in this strategy because OST believes they will address the most important environmental problems, accelerate the adoption and use of appropriate water quality standards, reduce burdens and impediments to program implementation, and promote broad participation in activities affecting the Nation's receiving waters. Each action is accompanied by milestones with the quarter and calendar year for completion. Work has begun on all of these actions. They have been organized in two groups: actions 1 through 6 are criteria-related actions, and actions 7 through 10 are standards-related actions, but are not arranged in any particular order. All ten actions are equally important; they are not listed in priority order.

Criteria-Related Actions

1. Issue implementation guidance for the 1986 bacteria criteria for recreation.

Milestones:

Publish §136 analytical methods for ambient water (final) .... completed, July 2003

Publish guidance (final) .... 1st Q, 2004

Publish §136 analytical methods for wastewater (proposed, final) .... 4th Q, 2004, 4th Q, 2005

This guidance is a major and immediate need due to the number of waters with bacteria impairments and the significant gaps in policy and technical guidance for implementing the recommended EPA criteria. It focuses on EPA's bacteria criteria published in 1986 for two bacterial indicators: *E. coli* and *enterococcus*. The guidance will assist states and authorized tribes with such issues as risk levels used in the criteria; implementation in NPDES permits, attainment decisions, and monitoring and advisories; and implementation in light of uncertainty inherent in the criteria. OST issued a draft of the guidance in 2002, and will publish the final guidance in 2004 after completing the review of comments and analysis of scientific information. Additionally, the guidance will assist states and authorized tribes that are required under the Beach Act of 2000 to adopt bacteriological criteria for coastal recreation waters that are as protective as EPA's criteria recommendations.

In the next two years OST will also publish approved analytical methods under 40 CFR part 136 for *E. coli* and *enterococcus*. States requested the methods to help measure attainment of the criteria and to support issuance of discharge permits.

---

2. Produce and implement a strategy for the development of pathogen criteria for drinking water and recreational use.

Milestones:

Develop a strategy for producing cryptosporidium
criteria for source waters .... 4th Q, 2003

Develop a strategy for revising existing criteria
for recreational waters .... 4th Q, 2003

Develop a strategy for establishing integrated
microbiological water quality criteria .... 3rd Q, 2004

Issue revised criteria document for recreational
waters (draft, final) .... 2nd Q, 2005, 2nd Q, 2006

Issue cryptosporidium criteria document
(draft, final) .... 4th Q, 2006, 4th Q, 2007

Issue integrated microbiological criteria document .... *(to be determined)*

According to the 2002 state section 303(d) listings, pathogens are the second most frequent cause of water quality impairments under the Clean Water Act. Increasing interactions between humans and domesticated and feral animals are increasing the incidence of human microbial disease and contributing to the evolution of new human pathogens. Some microbes that originally had animal hosts have acquired the ability to infect humans. A number of initiatives such as the Interim Enhanced Surface Water Treatment Rule and the 2000 Beach Act are important in reducing the risk of waterborne microbial disease and will continue. In light of emerging risks, OST, along with other participating OW offices, is developing a Strategy for Waterborne Microbial Disease Control. The microbial strategy will contain ongoing and needed actions selected by EPA technical work groups and reviewed by scientists and the public.

The milestones identified here are a component of the draft strategy for Waterborne Microbial Disease Control findings.

- Developing criteria for *Cryptosporidiumparvum*. At this time we do not know if there will be additional microbes regulated under the Safe Drinking Water Act that will require ambient water quality criteria for drinking water sources.

- Developing revised criteria for ambient water quality criteria for recreational waters, in accordance with the Beach Act of 2000 (which requires new or revised criteria by 2005), based on an assessment of potential human health risks resulting from exposure to pathogens in coastal recreation waters, and development of appropriate and effective indicators for the presence of pathogens that are harmful to human health.

- Development of water quality criteria that integrates protection against harmful exposures to pathogens for drinking sources and recreational waters, and will also consider health protection for other ambient water uses, e.g., shellfish growing.

Because of the complexity of the issues involved, the first step shown for each of the three needs above is to construct a specific strategy for criteria development. The three strategies will review available scientific studies and data, and assess various options for developing the criteria.

> **3. Produce and implement a strategy for the development of suspended and bedded sediment criteria.**
>
> Milestones:
>
> Consult EPA Science Advisory Board (SAB) .... 4th Q, 2003
>
> Issue the suspended and bedded sediment
> criteria strategy .... 2nd Q, 2004

Sedimentation and siltation problems account for more identified water quality impairments of U.S. waters than any other pollutant. Developing quantifiable water quality criteria for sedimentation will require research to identify sedimentation indicators, analytical methods, ecological relationships, reference conditions, and waterway classification systems. As a first step, OST will develop a strategy in 2003 for how best to develop such criteria. The strategy will set the course that will ultimately lead to suspended and bedded sediment criteria. OW's Office of Wetlands, Oceans and Watersheds (OWOW) has coordinated the development of guidance for TMDLs involving sediment, including an assessment of the state of knowledge and innovative guidance on assessing watersheds for river stability and sediment supply. Additionally, OST and OWOW are working with the Office of Research and Development (ORD) to pursue sedimentation research as part of ORD's aquatic stressors framework and implementation plan for effects research.

4. Provide technical support to states and tribes for developing and adopting nutrient criteria and biological criteria.

Milestones:

<u>Nutrient criteria:</u>

Assist in review of state plans for criteria development 2003–2004

Issue methods manual for wetlands .... 4th Q, 2003

Establish enhanced technical support process .... 1st Q, 2004

Issue criteria document for selected estuary
and coastal waters 2005–2007

Issue criteria document for selected wetland regions 2005–2007

<u>Biological criteria:</u>

Update survey of state and tribal programs .... 2nd Q, 2003

Issue methods for the use of statistics in bioassessments
and biocriteria development .... 4th Q, 2004

Issue methods for the use of bioassessments to
refine designated aquatic life uses .... *(see #7 below)*

Develop the scientific relationships between
bioassessments, biocriteria, chemical criteria
and other forms of criteria .... 4th Q, 2005

Issue coral reef methods .... 4th Q, 2006

Issue large river methods .... 4th Q, 2007

Issue stressor identification support system .... 4th Q, 2007

Issue Great Lakes methods .... 4th Q, 2008

Support implementation for streams, small rivers,
and other water bodies .... *Ongoing*

Nutrient-related issues also rank among the highest needs for the criteria program. Excessive nutrients are among the top four leading causes of water quality impairments. Most states recognize the need for such criteria, but because of the difficulty and complexity of the task, only two states to date have established a complete numeric baseline for nutrient problems and even these are specific to lakes only. In 2001–2002 OST issued 28 nutrient criteria documents covering all freshwater lake and river ecoregions, and guidance recommending that states establish plans for developing and adopting criteria. To date, 32 states have submitted nutrient criteria plans to EPA for comment and 9 additional states hope to submit plans this year. These nutrient criteria plans are expected to be developed in collaboration with EPA and include milestones and schedules for each state to work on the complex tasks of gathering and analyzing scientific data and adopting criteria into water quality standards. When developing nutrient criteria, a state or tribe has the flexibility to refine EPA's recommended approach and criteria to better reflect local conditions and data availability. In 2003, OST will work with states and tribes to review and revise EPA's technical support efforts to ensure they best support state needs.

Biological criteria and assessments are taking on increased importance in water quality programs. There is a growing recognition of the importance of biocriteria and bioassessment techniques in water quality protection and measuring the success of clean-up efforts. Biocriteria are particularly useful in advancing the scientific basis for designating aquatic life uses and can be an important tool for conducting use attainability analyses for aquatic life uses. They can also be used as an "ecological check" to see whether regulation of individual chemicals is achieving expected results. The National Research Council's 2001 report recommended that biological criteria be used in conjunction with physical and chemical criteria in Clean Water Act programs. The NRC recommends the expanded use of biocriteria because they are directly related to aquatic life designated uses, they are waterbody response criteria, and they integrate effects of multiple stressors over time and space. Biological criteria can also play an essential role in determining the highest attainable uses for aquatic life and in conducting more scientifically defensible use attainability analyses.

With active leadership from EPA and states, all states now have bioassessment programs for streams and small rivers, and over half the states have adopted at least narrative biocriteria into their water quality standards. Nevertheless, states and tribes need continued support to strengthen the use of biocriteria in water quality standards and to initiate the use of biocriteria for other water body types in addition to streams and small rivers. The milestones in this priority strategic action are designed to provide the products and support states and tribes have requested. OST and its partners will continue to provide technical support and assistance to states and tribes to complete adoption of biocriteria for streams and small rivers and to develop and adopt biocriteria for all other water body types. OST will continue to work to produce methods for developing biocriteria for all remaining water body types for which a method is currently not available, including large rivers, great lakes, intermittent and ephemeral headwaters and coral reefs.

DIRECTLY

▶In the early 1970s the academic community conceived the idea of systematically assessing local aquatic biology with field studies and quantitative biological criteria.

▶Several states (OH, MO, MI, NC, ME, NY) began testing and using this approach.

▶EPA has provided extensive technical guidance, policy recommendations and technical assistance.

▶All states have a bioassessment program for streams and small rivers, and for these waters:

   ▶29 states have adopted narrative biocriteria into water quality standards

   ▶23 states have quantitative translators for narrative biocriteria (8 more are under development)

   ▶4 states have adopted numeric biocriteria into water quality standards (9 more are under development)

Source: *Summary of Biological Assessment Programs and Biocriteria Development for States, Tribes, Territories and Interstate Commissions: Streams and Wadeable Rivers (EPA-822-R-02-048)*

5. Develop and apply a systematic selection process to produce new and revised water quality criteria for chemicals to address emerging needs.

Milestones:

Draft criteria selection process .... 1st Q, 2004

Implement final criteria selection process .... 3rd Q, 2004

OST agrees with stakeholders that there is an urgent need to develop new and updated water quality criteria. The growing need to keep abreast of emerging contaminants of concern as well as new information on familiar constituents is a constant challenge. So too are the rising costs of developing individual criteria documents. OST will work with partners to prioritize chemicals and develop new and revised criteria as rapidly as possible. The key to successful use of limited resources is to focus on developing those criteria that will have the greatest effect across the country, fill critical gaps, and reduce uncertainty in water quality management decisions. OST will establish a systematic process that takes these factors into account when selecting criteria for development and will then derive new and revised criteria based on this process.

6. Complete the national consultation with the Federal Services on existing aquatic life criteria.

Milestones:

Develop methodology for evaluating effects of pollutants on endangered and threatened species .... 2nd Q, 2004

Conclude biological evaluation for first batch of pollutants .... 3rd Q, 2004

Conclude biological evaluation for subsequent batches of pollutants .... *every 6 months*

Protection of threatened and endangered species is important in standards development. EPA, states and tribes have certain obligations under the Endangered Species Act to protect threatened and endangered species. EPA's obligations extend to consulting with the U.S. Fish and Wildlife Service (FWS) and the National Marine Fisheries Service (NMFS) whenever we approve state- or tribal-adopted water quality standards, and when we promulgate federal standards. States and tribes also have an obligation to consider Endangered Species Act concerns during the development of their water quality standards.

The national consultation on 49 aquatic life water quality criteria is a key action established in the 2001 memorandum of agreement between EPA, the FWS, and the NMFS regarding enhanced coordination under the Clean Water Act and Endangered Species Act. The consultation is particularly important because water quality standards containing these criteria are the basis for many TMDLs, permits and other actions. The first step in the consultation is for EPA to prepare biological evaluations of the degree to which each criterion may affect endangered and threatened species. A team of EPA and Service scientists has drafted a methodology for these evaluations. It will undergo peer review before being finalized and applied to review specific criteria.

The memorandum of agreement specifies other actions including consulting on new and revised standards and on certain NPDES permits, conducting cross-training between the agencies, organizing early participation of the three agencies in triennial reviews of water quality standards, and elevating unresolved issues to management's attention. Most of these activities are currently underway.

Standards-Related Actions

7. Provide technical support, outreach, training and workshops to assist states and tribes with designated uses, including use attainability analyses and tiered aquatic life uses.

Milestones:

Develop a plan for providing outreach, training, workshops and other support for states and tribes on critical issues regarding designating appropriate uses .... 2nd Q, 2004

Issue methods for the use of bioassessments to refine designated aquatic life uses .... 4th Q, 2004

Internet/Web-based clearinghouse operating with information supporting establishment of designated uses .... 4th Q, 2004

Clean Water Act regulatory programs, such as discharge permits and TMDLs, are geared toward achieving water quality standards. The public relies on EPA, the states and authorized tribes to set designated uses that reflect the goals of the Clean Water Act. This priority strategic action will help clarify states and tribes understanding of how to conduct use attainability analyses (UAAs). It will help states and tribes to make decisions related to adjustments of uses such as when higher uses can

be attained but are not designated in standards or when higher uses have been designated that cannot be attained. Additionally, this action will help states and tribes decide when use adjustments should not be made, such as removing a designated use that is being attained, has been attained since 1975, or can be attained.

Providing this support will fill a major program gap, promote more efficient use of resources, and ultimately lead to incorporating the highest attainable uses into water quality standards. States consistently rank this as the single most urgent need from EPA. Some participants believe that lack of clarity from EPA on designated use issues has prolonged local debates over the ultimate goals for water bodies and has resulted in a stalled clean-up progress in the meantime. OST will work with other EPA offices, states, authorized tribes and other partners to help resolve use-related issues, such as how natural conditions, or irretrievable human-caused conditions, or economic factors may be considered, and what types and quantity of data are needed for use attainability analyses.

The National Research Council's 2001 TMDL report said that "assigning tiered designated uses is an essential step in setting water quality standards." OST does not agree tiered uses are essential for all situations, but does agree that refined uses including biologically "tiered" uses can improve the effectiveness and credibility of state and tribal standards in many situations. Broad uses such as "Fish and wildlife use" or "Recreational use" are fully acceptable under the Clean Water Act, although EPA and many states are learning that refined uses offer advantages for waters where information is available to develop them. For example, they can provide better operational definitions of desired outcomes, and can provide flexibility to describe locally-important variations that broad uses may not. For aquatic life uses, OST is developing methods to show how biological criteria can help inform the adoption of highest attainable uses. Further, OST is developing biological criteria tools that show how the degree of human disturbances in a watershed can affect ecological outcomes. Many states have been using biological assessments and biological criteria in their standards to protect high quality waters and provide goals for improving degraded waters. OST will work with ORD, OWOW and other partners to develop methods that will help states and tribes understand the benefits and scientific rationale behind bioassessment-supported designated uses for aquatic life.

> **8. Provide implementation support concerning technical issues affecting permits and TMDLs, beginning with technical support and outreach concerning the duration and frequency component of existing water quality criteria.**
>
> Milestones:
>
> Provide support for duration and frequency component
> of existing water quality criteria .... 4th Q, 2005
>
> Provide support for mixing zone policies .... 4th Q, 2005
>
> Provide support for additional technical issues .... *Ongoing*
>
> Develop implementation methods for new water
> quality criteria as needed *(see #5 above)* .... *Ongoing*

Water quality standards and criteria provide the environmental baselines needed to regulate discharges to water and determine the extent of clean-up actions. New collaboration across programs must occur to solve the Nation's water quality problems. In particular, there must be a common understanding of the how standards and criteria will be applied. Modifying criteria on a site-specific basis and applying the criteria for specific purposes often involve complex assumptions about pollutant fate and transport, mixing zones, pollutant sources, fluctuations in discharge rates and receiving water flows and chemistry, and biological processes.

The goal of this priority strategic action is to enable states and tribes to implement criteria effectively, considering the scientific basis, in monitoring design, attainment decisions, TMDL development, site-specific conditions, and permit issuance. OST and its partners will provide technical support, training and outreach for implementing the duration and frequency components of existing numeric criteria, and in establishing and applying mixing zone policies. Additionally, OST will provide technical support, training and outreach on additional implementation issues of importance (e.g., wet weather). On an ongoing basis, OST with its partners will also develop new implementation support or reference appropriate existing implementation guidance when issuing new or revised criteria documents.

MANY IMPORTANT ACTIONS ARE LINKED TO STANDARDS

- Assessing which U.S. waters are impaired and not impaired.
- Establishing targets and load reductions needed in impaired waters through TMDLs.
- Setting limits on pollutants discharged through enforceable NPDES permits.
- Issuing permits for dredge or fill activities.
- Certifying that other federal licenses or permits comply with standards.
- Establishing applicable or relevant and appropriate requirements for on-site responses at Superfund sites.

9. Identify any drinking-water source waters whose water quality standards do not protect the use, and work with regions, states, and tribes to correct any deficient standards as soon as possible.

Milestones:

Letters to states requesting that they review drinking water use protection in their water quality standards .... 4ᵗʰ Q, 2003

Geographically-referenced information available to track progress toward this goal .... 4ᵗʰ Q, 2005

In September 2000, states reported that there were approximately 180 million people served by public drinking water systems using surface water sources—rivers, streams, lakes and reservoirs. Under the Safe Drinking Water Act, states are mandated to assess each of their source waters in order to determine the susceptibility of public water systems to threats in their watersheds. These assessments will help to protect source waters more effectively and prevent pollutants from entering the waters in concentrations harmful to human health. The Clean Water Act will play a major role in these efforts and includes many regulatory and non-regulatory tools that can protect source waters. Full use of those tools can only occur, however, if the water quality standards for those waters are fully protective. OST, along with EPA's Office of Ground Water and Drinking Water (OGWDW) and EPA's regional offices, will work with states and authorized tribes to identify and correct any state water quality standards that do not provide adequate protection for contaminants of concern for drinking water usage. For example, in 2003 we will ask states and tribes to work with EPA to identify any drinking water intakes located in source waters that have not been designated for public water supply uses or do not have equivalent protections in place to protect the intakes. OST will also work with other EPA offices to draw on information in geographically-referenced databases containing intake locations and water quality standards to establish a way of reporting progress to this goal by 2005. OST and the regions will address any remaining issues in carrying out EPA's oversight functions.

> **10. Develop a web-based clearinghouse for exchanging information on critical water quality standards issues, beginning with antidegradation.**
>
> Milestones:
>
> Establish a test web site with pages for state and tribal review containing sample antidegradation information .... 4th Q, 2003
>
> Clearinghouse operating with information supporting development of state and tribal antidegradation programs ....2nd Q, 2004
>
> Internet/web-based clearinghouse operating with second round of information supporting establishment of designated uses .... 4th Q, 2004

Several stakeholders suggested during listening sessions that EPA should establish a means for sharing information about approaches that have worked for some states and tribes and could potentially be applied elsewhere. The suggested "clearinghouse" or "resource center" approach has been a successful way to share information in other programs. The clearinghouse should be accessible to all who could benefit from the information. Ideally EPA would play an active role in seeking materials and providing assistance in using them. A clearinghouse would be particularly useful for emerging issues where a few states or tribes have had success in specific areas and where discussions between EPA, other states and other tribes could foster creative solutions.

OST will be developing this clearinghouse with an initial focus on antidegradation, since stakeholders indicated the importance of addressing antidegradation. EPA's regulation requires states and authorized tribes to adopt antidegradation policies and to identify implementation methods for the policies. Antidegradation procedures are designed to preserve water quality in outstanding water resources; keep clean waters clean where possible, considering important social and economic development; and prevent loss of existing uses through degradation. Implementing such procedures can prevent further waters being added to the list of impaired waters needing TMDLs. Several stakeholders and commenters indicated that the most important immediate need is for sharing of information about antidegradation requirements and implementation methods. In the absence of such a central source of information, each state and tribe would need to independently develop its own approach without being able to learn from the successes and experiences of other states and tribes who have already gone through the process. The clearinghouse will also assist OST and the regions to provide ongoing technical support and outreach on important antidegradation issues.

OF DRINKING WATER

▶180 million people use 14,136 public water systems that are supplied by surface water.

▶The pesticide atrazine has been detected in over 90% of Ohio's public surface water systems and in similar percentages elsewhere.

▶Concentrated animal feeding operations are believed to be among the major sources of microbial pathogens in drinking water.

▶Conventional drinking water treatment systems are not fully effective for all pathogens and are ineffective for most pesticides like atrazine.

Next Priority Strategic Actions

The five strategic actions below constitute the next set of priorities that will continue to guide OST activities presently and in the coming years. Many of these "next priority" actions already have activities and workplans underway, while others are in the planning stages and do not yet have milestones established. The actions identified here are also fundamentally important to the advancement of clean water goals. The designation as "next priority" reflects OST's commitment to these priority activities which are outstanding needs but will be delivered over a longer time.

> **1. Update the aquatic life methodology for developing ambient water quality criteria.**

Improved methodologies for criteria will enable future criteria to address important toxicological endpoints and exposure routes appropriately, and will help develop future criteria that can be used with refined designated uses. For aquatic life protection, EPA scientists and non-EPA stakeholders agree that EPA's 1985 guidelines for deriving numeric national aquatic life criteria require updates and refinements to reflect advances in scientific understanding and the increased complexity of water quality problems. The 1985 guidelines are not preventing development of scientifically appropriate criteria, but they lack specificity to address emerging needs efficiently. OST and ORD, with assistance from FWS, will collaborate to update the guidelines in a priority sequence with interim products.

> **2. Provide technical support, outreach, and training to assist states and tribes implementing mercury criteria in assessments, TDMLs, and permits.**

Mercury contamination is the leading cause of public advisories concerning allowable quantities of fish to eat. In 2001, EPA published a new water quality criterion for methyl mercury in fish tissue for the protection of human health. In publishing the criterion, EPA recognized that there are important issues relating to implementing the criterion in regulatory programs. OST has established an EPA technical workgroup to develop information and approaches for states and tribes to implement the recommended criterion. This group is exploring options for deriving water quality-based effluent limitations and TMDL target values from the EPA mercury criterion expressed as fish tissue contamination levels. The draft is expected in late 2003.

> **3. Provide technical support, outreach, and training to assist states and tribes in refining human health criteria to reflect local bioaccumulation and fish consumption patterns.**

The *Methodology for Deriving Ambient Water Quality Criteria for the Protection of Human Health*, issued by OST in 2000, included improved consideration of exposure routes and toxicological endpoints. The methodology includes new protocols for fish consumption rates and bioaccumulation that can vary considerably depending upon local conditions. OST plans to develop a technical support document (TSD) entitled *Technical Support Document Volume 2: Development of National Bioaccumulation Factors*. Additionally, OST plans to publish a detailed version of the national bioaccumulation methodology included in the 2000 *Methodology* and another TSD to provide methods for deriving site-specific Bioaccumulation factors (BAFs). OST will work with OGWDW to harmonize criteria regarding surface water pollutants that are of concern for drinking water supplies.

> **4. Provide updated analytical methods for new and existing criteria.**

Initial emphasis will be on methods for measuring metals and other pollutants that appear most frequently in NPDES permit limitations. OST will develop methods for emerging pollutants on a priority basis as needed, including a method for polybrominated diphenyl ethers (PBDEs). Additionally, OST will promulgate a final rule in 2003 that will make available analytical methods for bacteria (*E. coli* and *enterococci*) and protozoa (*Cryptosporidium* and *Giardia*) in ambient water. OST is also validating analytical methods for *E. coli* and *enterococci* in effluents during 2003 to make these methods available for NPDES permits and TMDL monitoring. OST is also investigating other pathogens and plans to validate methods for *Cryptosporidium* in effluents to be available when a *Cryptosporidium* water quality criteria is issued.

> **5. Foster broad participation in the setting of water quality standards by providing training, outreach, and education, including Internet-based distance learning access to the Water Quality Standards Academy.**

As clean water benefits all Americans, water quality standards are essential for clean water protection. When standards were first being set decades ago, participation centered on EPA and state technical experts and a few

---

RECOGNIZE POLITICAL BOUNDARIES

Of the 2,165 watershed sub-basins in the lower 48 states:

▶ Almost all cross county lines

▶ 667 (31%) contain parts of two or more states

▶ 247 (11%) contain Indian reservations

▶ 64 (3%) are shared with Canada or Mexico

---

interested stakeholders at the state and national level. As water quality issues become more prominent, more participants from broad sectors are becoming interested and involved in water quality standards issues. EPA, states and tribes today increasingly work with other federal, state, tribal and regional government agencies, the regulated community, a wide variety of economic sectors, water resource agencies, and private citizens. Interactions on issues can occur at the statewide or reservation-wide scale, as well as locally or watershed-wide.

To support and encourage these trends, OST will work with other internal EPA offices, other Federal Agencies and external organizations to better educate and inform EPA's partners, stakeholders and the public about water quality standards and the role these groups can take in the standards setting process. OST will use printed and visual media, the Internet, conferences and workshops, and state-of-the-art distance learning mechanisms to communicate information and provide a limited number of face-to-face training sessions. OST will continue to offer the Water Quality Standards Academy to provide training and will upgrade this popular training course to a web-based environment to ensure greater access to program information. As knowledge about the program increases, OST will provide more advanced, in-depth training and will expand outreach activities to include broader audiences. It is OST's expectation that better informed and educated citizens will result in greater involvement and participation in the water quality standards setting process at the local or watershed level.

OST will focus much of this outreach and communication on participants in watershed planning and protection. A 2002 OW study, *A Review of Statewide Watershed Management Approaches, Final Report,* found that the water quality standards development process is not significantly involved in the watershed management approaches of eight states studied, but rather occurs primarily on a statewide basis. Several states indicated, however, that the statewide watershed approach has indirectly benefitted the water quality standards process by improving the level of communication about standards among state partners, increasing public understanding and enhancing the state's ability to assess the need for revisions.

Implementing the Strategy

Roles of EPA Offices and Key Partners During Implementation

The EPA offices with primary responsibility for the water quality standards and criteria program are OST and EPA's ten regional offices. Other EPA offices play important roles in developing and implementing water quality standards, including OW offices responsible for monitoring, assessments, TMDLs, permits, wetlands, oceans, and drinking water, as well as ORD and the Office of General Counsel. OST will establish additional work groups with representatives from the regions and these other offices (as well as the Office of Policy, Economics and Innovation, or OPEI) to implement the priority strategic actions in this strategy in the timeframes provided, barring any unforeseen events.

The ten EPA regional offices have an important and special role in the water quality standards and criteria program. OST will work with its regional counterparts to develop a collaborative system for administering the water quality standards program, including but not limited to the priority strategic actions. The system should recognize geographic and ecological differences and still maintain minimum requirements and certain levels of consistency nationwide. For example, OST staff could generally focus on issues having national significance while EPA regional offices could take the lead on local, site-specific issues. Additionally, EPA regional offices can help integrate water quality monitoring with water quality standards activities, including using environmental information to help target standards actions and assisting in correctly interpreting standards when making attainment and permitting decisions. Examples of important activities undertaken by EPA regional offices include serving as liaisons to states and tribes; helping states and tribes develop additions and revisions to their standards that are consistent with federal requirements and address high-priority needs; providing advice where needed on specific standards development and implementation issues; developing criteria methods for pollutants affecting regionally-important waters; guiding priorities for triennial reviews; reviewing and approving new and revised water quality standards; and coordinating with the regional and district offices of the FWS and the NMFS regarding endangered and threatened species issues.

Implementing this strategy will also require greater coordination and cooperation between EPA and key external partners than in previous years. Recent cooperative efforts should continue, adjusting for lessons learned in the process. Current efforts include:

- Meeting with states and tribes on a regular basis to oversee the directions of the program, through such groups as the State/EPA Operating Committee, the Tribal Operations Committee, the State/EPA Workgroup on Water Quality Standards, the Federal/State Toxicology and Risk Assessment Committee, and other fora. These groups can help EPA review implementation of the strategy and provide valuable feedback.

- Obtaining state and tribal input on operational issues before releasing important technical support documents.

- Working in watershed-based partnerships to develop and share information for developing standards and criteria. For example, a broad range of organizations cooperated in efforts to develop regional criteria guidance for dissolved oxygen, water clarity and chlorophyll for the Chesapeake Bay and its tidal tributaries.

- Using Regional Technical Assistance Groups when developing and implementing EPA's recommended criteria for nutrients. These groups, consisting of technical staff from EPA regions and states as well as other researchers, work at the regional level to assemble environmental data and develop analytical approaches. In the future, these groups will become more involved in implementation issues as states and authorized tribes develop nutrient criteria plans and adopt nutrient criteria.

- Using an EPA technical workgroup to help develop implementation methods for EPA's recommended criteria for methylmercury. This group is identifying questions that need answering for deriving water quality-based effluent limitations and TMDL target values from the EPA mercury criterion expressed as fish tissue contamination levels.

- Using quality-assured data generated by non-governmental parties where possible and appropriate for development of water quality criteria. EPA works with these groups to ensure that data adhere to EPA protocols. Also, EPA retains the governmental responsibility to establish the protocols, review the results, conduct peer review, and issue the criteria as federal recommendations.

OST will continue to engage the scientific community and the public in criteria and technical support development. Specifically, we will continue the practice of notifying the public when starting a new or revised criteria or guidance document and of seeking scientific data and information at various stages of criteria development. We will also continue to seek peer review of resulting criteria and simultaneously make them available for further scientific input from the public. This approach will help EPA publish water quality criteria reflecting the latest scientific knowledge.

Additionally, OST will utilize open public processes wherever possible. For example, OST may use public symposia, meetings of professional societies and other open venues to obtain information and ideas for technical support documents. OST will also continue to coordinate EPA-sponsored research activities consistent with the priority strategic actions in this strategy.

Future Strategy Refinements

OST is now working with its partners to implement the priority strategic actions in this strategy. When implementing the strategy, OST will stay attuned to the needs of its partners. For example, OST will from time to time request feedback concerning how well the strategy is succeeding. Products will be available on EPA's web site at <u>http://www.epa.gov/waterscience</u>. As implementation experience grows, OST may revise the strategy as determined by need over time (applying the strategic themes as issues emerge) to continue the selection of priorities. When revising the strategy, OST will again seek input from our many partners and the public. Ten items are presented in Attachment 2 that OST and subsequent workgroups may decide to designate as priorities at a later date.

Other Ongoing Program Activities

Notwithstanding the priority strategic actions in this strategy, OST will continue to perform other core functions in support of water quality standards and criteria. Many of these functions are mandated by the Clean Water Act, the Endangered Species Act, or other statutes. These include:

• Oversight of national water quality standards actions: overseeing water quality standards development, reviewing draft state or tribal standards, recommending improvements, reviewing new and revised standards, approving or disapproving them, and issuing federal replacement standards to correct deficiencies where necessary.

• Endangered Species Act obligations: consulting with the FWS and the NMFS on federal actions that may affect endangered and threatened species, and carrying out obligations pursuant to biological opinions from the Services.

• Coordination with research activities: coordinating with EPA's ORD to ensure that the most evolved and advanced scientific research is available to support water quality standards and criteria.

• Technical assistance: providing case-by-case guidance, technical assistance, data and information, and referrals to regional, interstate, state, tribal and local water quality managers undertaking program activities such as standards development, TMDL development, permitting, monitoring and modeling, among others.

• Public access: providing electronic access to state, tribal and federal water quality standards, including displaying adopted and approved designated uses and criteria in nationally comparable tabular and map form for all waters of the United States in the Water Quality Standards Database at http://www.epa.gov/waterscience/standards/states/.

• Program tracking and reporting: tracking progress made by states and tribes in adopting and revising standards. Reporting and managing activities under the Government Performance and Results Act and the Federal Managers Financial Integrity Act.

Conclusion

STRATEGY FOR WATER QUALITY STANDARDS AND CRITERIA

Water quality standards and criteria are the foundation of water quality protection programs under the Clean Water Act and the Safe Drinking Water Act. Water quality standards and criteria issues impacting the programs such as assessments, TMDLs and permits are increasingly complex. The priority strategic actions in this strategy will strengthen the foundation of water quality programs, fill critical gaps and implementation needs, help deal with uncertainty and complexity, and ultimately assist in attaining clean water goals.

Carrying out the strategy will require joint efforts among EPA and its partners and will also entail creativity and new approaches. Partners will have key roles in developing products and implementing the work outlined in the strategy. As implementation continues, OST may periodically make mid-course corrections to keep the strategy current and focused.

Acronyms

ASIWPCA means the Association of State and Interstate Water Pollution Control Administrators.

EPA means the U.S. Environmental Protection Agency.

FWS means the U.S. Fish and Wildlife Service (FWS) in the U.S. Department of the Interior.

NMFS means the National Marine Fisheries Service (NMFS) in the National Oceanic and Atmospheric Administration of the U.S. Department of Commerce.

NPDES means the National Pollutant Discharge Elimination System, established by section 402 of the Clean Water Act.

OGWDW means EPA's Office of Ground Water and Drinking Water.

ORD means EPA's Office of Research and Development.

OST means the Office of Science and Technology in EPA's Office of Water.

OW means EPA's Office of Water.

OWOW means EPA's Office of Wetlands, Oceans and Watersheds.

TMDL means total maximum daily load. States develop total maximum daily loads for certain water bodies that do not attain applicable water quality standards. See section 303(d) of the Clean Water Act.

Strategy for Water Quality Standards and Criteria

Acknowledgments

This strategy was developed by a workgroup consisting of Fred Leutner (workgroup leader), EPA Office of Science and Technology, Washington, DC; Heidi Bell, EPA Office of Science and Technology, Washington, DC; Libby Chatfield, West Virginia Environmental Quality Board, Charleston, WV; Linda Holst, EPA Region 5, Chicago, IL; Catherine Kuhlman, EPA Region 9, San Francisco, CA; Cara Lalley, EPA Office of Science and Technology, Washington, DC; Terry Oda, EPA Region 9, San Francisco, CA; Joseph Piotrowski, EPA Region 3, Philadelphia, PA; Deborah Smith, California Regional Water Quality Control Board, Los Angeles, CA; Anthony Maciorowski, EPA Office of Science and Technology, Washington, DC; and Scott Ireland, EPA Office of Science and Technology, Washington, DC.

OST wishes to thank the following organizations for participating in the development of the strategy: water quality managers and water quality standards experts from the states of Arizona, California, Connecticut, Florida, Illinois, Kansas, New York, Oklahoma, Pennsylvania, South Carolina, Texas, Utah, Washington and West Virginia; the Association of State and Interstate Water Pollution Control Administrators; Water Division Directors and program staff in EPA Regions 1 through 10; EPA headquarters program managers for the NPDES program, TMDL program, Safe Drinking Water Act programs, wetlands programs, oceans programs and water law counsel; the Federal-State Toxicology and Risk Assessment Committee; the Federal Water Quality Coalition; the Water Environment Federation; the Association of Metropolitan Sewerage Agencies; the Clean Water Network and its participating members; the Utility Water Act Group; and the Electric Power Research Institute.

Information Sources for this Strategy

States

State water quality managers and water quality standards experts, particularly those from the states of Arizona, California, Connecticut, Florida, Illinois, Kansas, New York, Oklahoma, Pennsylvania, South Carolina, Texas, Utah, Washington and West Virginia.

EPA

EPA water program staff, including directors, managers and staff with responsibility for water quality standards, water quality monitoring and assessments, TMDLs, NPDES permits and drinking water in each of EPA's ten regional offices.

EPA Water Quality Standards Coordinators.

EPA program managers in headquarters for the NPDES program, TMDL program, Safe Drinking Water Act programs, wetlands programs, oceans programs and water law counsel.

Meeting of EPA headquarters and regional TMDL, NPS and assessment/monitoring coordinators, Albuquerque, NM, June 4-7, 2001.

Stakeholders, State Program-Specific Groups and Other Input

ASIWPCA TMDL Conference (Regions 5, 6 and 7), Austin TX, April 18–20, 2001.

Meeting with Federal-State Toxicology and Risk Assessment Committee, May 22, 2001.

Meeting with Federal Water Quality Coalition, May 30, 2001, June 28, 2001.

Conference call with State/EPA TMDL Coordinators, July 13, 2001.

Meeting with Water Environment Federation, July 17, 2001.

Meeting with Association of Metropolitan Sewerage Agencies, August 23, 2001.

Meeting with Clean Water Network, August 28, 2001.

Letter from American Fisheries Society, September 13, 2001.

Letter from Clean Water Network, September 21, 2001.

References

Association of State and Interstate Water Pollution Control Administrators, *A State Perspective: Future Needs/Directions of the WQS Program.* Distributed at the mid-year meeting, March 22, 2001.

EPA, Office of Water, *Water Quality Criteria and Standards Plan—Priorities for the Future*, interim final. June 1998, EPA 822-R-98-003.

EPA, Office of Water, Advance *Notice of Proposed Rulemaking, Water Quality Standards Regulation.* July 7, 1998, 63 FR 36741. Includes associated public record of written comments, and notes of discussions at public meetings.

EPA, Office of Water, *Protecting Public Health and Water Resources in Indian Country: A Strategy for EPA/Tribal Partnership.* October, 1998. Strategy for water quality standards is addressed on pp. 11–12.

EPA, Office of Inspector General, Central Audit Division, *Proactive Approach Would Improve EPA's Water Quality Standards Program.* Report No. 2000-P-001385-00023, September 29, 2000.

EPA, Office of Science and Technology, *An Assessment of the Water Quality Standards Development and Review Process*, Final Report, October 2000.

EPA, Office of Water, *A Review of Statewide Watershed Management Approaches*, Final Report, April 2002. http://www.epa.gov/owow/watershed/approaches_fr.pdf.

EPA, Stage 2 Microbial and Disinfection Byproducts Federal Advisory Committee, *Agreement in Principle,* September 2000, as published in 65 FR 83015, December 29, 2000.

EPA, Office of Science and Technology, *Perceptions on TMDL Technical Support: Input from State, EPA, Discharger Organizations, and Clean Water Action Network*, December 29, 2000.

EPA, Fish and Wildlife Service, National Marine Fisheries Service, *Memorandum of Agreement Between the Environmental Protection Agency, Fish and Wildlife Service and National Marine Fisheries Service Regarding Enhanced Coordination Under the Clean Water Act and Endangered Species Act.* February 22, 2001, 66 FR 11202.

EPA, Advisory Committee on Water Information, *TMDL Science Issues Conference 2001: Closing Session Summary*, March 7, 2001.

EPA, Office of Water, *Guidance: Coordinating CSO Long-Term Planning with Water Quality Standards Reviews,* July 31, 2001, EPA-833-R-01-002

EPA, Office of Water, *Developing Strategy for Waterborne Microbial Disease*, August 29, 2001.

EPA, Office of Wetlands, Oceans and Watersheds, February 2, 2002, *A Review of Statewide Watershed Management Approaches,* executive summary in draft.

EPA, National Environmental Justice Advisory Committee, Fish Consumption Workgroup, draft reports, March 2002.

General Accounting Office, *Water Quality: Improved EPA Guidance and Support Can Help States Develop Standards That Better Target Cleanup Efforts,* GAO-03-308, February 2003.

National Research Council, Water Science and Technology Board, *Assessing the TMDL Approach to Water Quality Management*, June 22, 2001.

Other Sources of Information:

Regular meetings of the State/EPA Operations Committee.

Regular meetings of the workgroup on Water Quality Standards.

Regular meetings of the EPA Tribal Operations Committee and the Tribal Caucus of the committee.

Strategic Actions to be Considered for Future Priority Setting

- Develop default bioaccumulation factors for use in developing water quality criteria for the protection of human health. EPA's *Methodology for Deriving Ambient Water Quality Criteria for the Protection of Human Health*, October 2000, incorporates a number of scientific advancements, one of which is the assessment of exposure to humans through the food chain pathway. For bioaccumulative chemicals the food chain pathway is more important than ingestion of water. To assess exposure to bioaccumulative chemicals, EPA's methodology emphasizes the use of a bioaccumulation factor (BAF), which accounts for chemical accumulation in fish and shellfish from all potential exposure routes. EPA is currently finalizing a Technical Support Document that presents the technical basis for the national approach to developing BAFs. Because of the need for local data, not all states and tribes would have the ability to develop BAFs with limited resources. To address this concern, EPA could derive national default BAFs for specific bioaccumulative pollutants. It is envisioned that states and authorized tribes would use the national default BAFs as a starting point in the process of deriving appropriate and applicable water quality standards. The supporting literature searches and data analyzed in the process of deriving national default BAFs could also serve as a valuable resource for deriving regional or site-specific BAFs.

- Conduct research on methods to assess risks of multiple stressors to wildlife populations. ORD, OST and other EPA offices could pursue research to fill the important need of assessing risks to aquatic-dependent and terrestrial wildlife. ORD's current work in this area is outlined in the recently-completed Aquatic Stressors Research Framework.

- Conduct research on chemical-specific criteria for wetlands. ORD, OST and other EPA offices could pursue research to fill the important need of assessing effects of contaminants on wetlands. ORD's work in this areas is outlined in a recently-completed Aquatic Stressors Research Framework.

- Provide technical support and outreach for making scientifically valid site-specific modifications of criteria. Such technical support and outreach, for example, would help states and tribes protect endangered and threatened species, and human populations who consume higher quantities of fish and shellfish. Site-specific tools will also assist states and tribes in refining criteria at the time they are refining designated uses.

- Promote increased use of ecological criteria and watershed-scale indicators as measures of healthy water bodies. Combining elements of chemical, physical and biological criteria in ecological risk evaluations can help define "ecological criteria" as measures of healthy water bodies. Such criteria and indicators have the potential of estimating the total response of a water body to potential alterations and stressors and identifying the appropriate scale for remediation, e.g., remediation in the stream along the riparian corridor or watershed-wide. Once ecological indicators are established for a water body, landscape-scale stressor-response relationships can be determined and used as a basis for the development of watershed-scale indicators and as predictive tools for watershed management. These new scientific tools could help states and tribes make water quality standards more ecologically-based and could set the stage for better watershed management. OST could focus on the integration of traditional criteria into ecological criteria. ORD could research and develop watershed-scale indicators and indices of watershed integrity. As useful approaches emerge, OST and ORD would develop case studies to illustrate how ecological criteria and watershed indicators work and would develop methods to assist states and tribes with their own implementation.

- Provide technical support and outreach to states and tribes on antidegradation implementation procedures. Antidegradation procedures are designed to preserve water quality in outstanding water resources, keep clean waters clean where possible, and prevent loss of existing uses through degradation. Implementing such procedures can prevent further waters being added to the list of impaired waters needing TMDLs. Many participants identified lack of explicit guidance on antidegradation implementation procedure as a major program gap. In the absence of such guidance, each state and tribe must independently develop its own approach with little certainty that EPA will approve it. OST could begin by improving the distribution of recent policies and state-specific decisions affecting antidegradation implementation, and will follow by focusing new technical support and outreach on important antidegradation issues.

- Review and update the 1994 WQS Handbook. This update could incorporate new policies and technical support issued since the *Handbook* was last published in 1994. It could also include a checklist of required standards elements. The *Handbook* could be issued in CD-ROM and/or online versions with hyperlinks to supporting materials.

- Develop a broad strategy for addressing inter-jurisdictional differences in water quality standards on shared waters. Recent listings of impaired waters have highlighted some differences in standards and interpretations of standards at state lines. OST could develop a strategy to address this issue. OST could work with other EPA offices, states and tribes to explore administrative and policy steps that could lead to a more systematic treatment of these issues.

- Obtain early EPA, FWS, NMFS involvement in state and tribal reviews of standards. Issues concerning endangered and threatened species have often slowed EPA's review of submitted standards. These problems could be minimized if the agencies could agree on the right approaches before states and tribes start to review and revise their standards. Under the 2001 Memorandum of Agreement, the Services agreed to participate in meeting with EPA and the states and tribes to discuss the extent of upcoming water quality standards reviews. EPA agreed to take the lead to schedule such meetings near the start of the triennial review process. OST could work with other EPA offices to help facilitate this early involvement, and to guide and support states and authorized tribes in adopting criteria to protect listed species.

- Expand on-line services and databases. Participants encouraged OST to develop more EPA web sites such as those containing all state and tribal water quality standards effective under the Act and those with interactive geographic information systems that link state and tribal standards to individual water bodies.

www.ingramcontent.com/pod-product-compliance
Lightning Source LLC
Chambersburg PA
CBHW080651180526
45168CB00008B/3379